Bibliographic information published by the German National Library:

The German National Library lists this publication in the National Bibliography; detailed bibliographic data are available on the Internet at http://dnb.dnb.de .

Imprint:

Print and binding: Books on Demand GmbH, Norderstedt Germany
ISBN: 9783346157560

This book at GRIN:

https://www.grin.com/document/539319

Paul Chacha

Assessment and Control of Operational Loss of Sugar During Sugar Cane Processing

Kagera Sugar Limited (KSL) Case Study

GRIN Verlag

ASSESSMENT AND CONTROL OF OPERATIONAL LOSS OF SUGAR DURING SUGAR CANE PROCESSING. KAGERA SUGAR LIMITED (KSL) CASE STUDY

PAUL CHACHA

Kagera Sugar Limited (KSL), Misenyi, Bukoba Tanzania

Keywords: *sugar production, operational sugar loss, molasses exhaustion, bagasse, sugar boiling house, sugar mill*

Abstract

This paper concentrated on assessing and controlling the operational loss of sugar in various sugar byproducts during cane processing in cane sugar factory. In the study, factors associated with operational loss were discussed and Kagera Sugar Limited (KSL) was taken as a case study. Operational sucrose loss is divided into two major categories namely, determined losses and undetermined losses. Determined losses are those which can be easily quantified through lab analysis, such as loss in molasses, filter cake and bagasse. Unlike determined losses, undetermined losses are difficult to estimate directly but rather by factory material balance. Undetermined loss is further classified into chemical loss, mechanical loss and administrative loss. In both losses, the loss through molasses is by far the largest of total loss encountered during cane processing. Objective of this paper was to develop means of increasing sucrose recovery by monitoring sugar loss. After analyzing data collected from KSL production reports, for crushing seasons 2015/16 and 2016/17 the operational loss was 2.83 and 2.87 pol % cane crushed in respective seasons. This means KSL loses significant amount of sucrose as compared to 2.20 pol% cane crushed for plants installed after 1994 as reported by (Sharma Chetan and Kumar Vikas, 2015). Sugar loss in molasses and bagasse was high than others, it was concluded that, the use of three laws of sucrose loss in molasses coupled with optimization of boiling house operations would result in minimized loss of sucrose in molasses. Also, improving extraction would result in minimum sucrose loss in bagasse.

Table of Contents

1. Introduction

1.1 Sugar production process

The raw materials used for sugar production at commercial scale are sugar cane and sugar beet because they produce enough sugar per hectare (Perk, 1973). In Tanzania in which KSL is exclusively an example, sugar is produced from sugar cane. Cane typically consists of 15% dissolved substances in which about 13% is sucrose and 2% are other substance such as other sugars i.e. fructose and glucose (0.9%), inorganic salt/ash (0.6%) and organic salt, organic acids, polysaccharides, gums, protein, amino acid and amides (0.5%). Cane also consists of insoluble matters which are vegetable fibers (13%) and soil (2%), and water (70%) (SMRI, 2012). A conventional cane sugar factory has a one intention, to extract about 87% of cane dissolved substances.

The cane sugar production process consists of cane size reduction, juice extraction (cane milling), juice clarification, filtration, evaporation, crystallization, centrifuging, drying and packing. Steam generation using bagasse as fuel for electricity generation and/or prime movers driving is also an important operation in sugar producing factories. See sugar production flow diagram presented in figure 1:

The purpose of any cane sugar factory is to maximize recovery of sucrose from cane and to produce high amount of sugar of the desired quality at minimum cost of production. However, from processing point of view, not all sucrose contained in sugar cane is recovered from the cane as commercial sugar. Some of it is lost in the course of cane processing in the factory. The major losses in a typical cane factory are due to Molasses, Bagasse and filter cake and undetermined factors. It's a job of production team to ensure that operational loss is kept as minimum as possible.

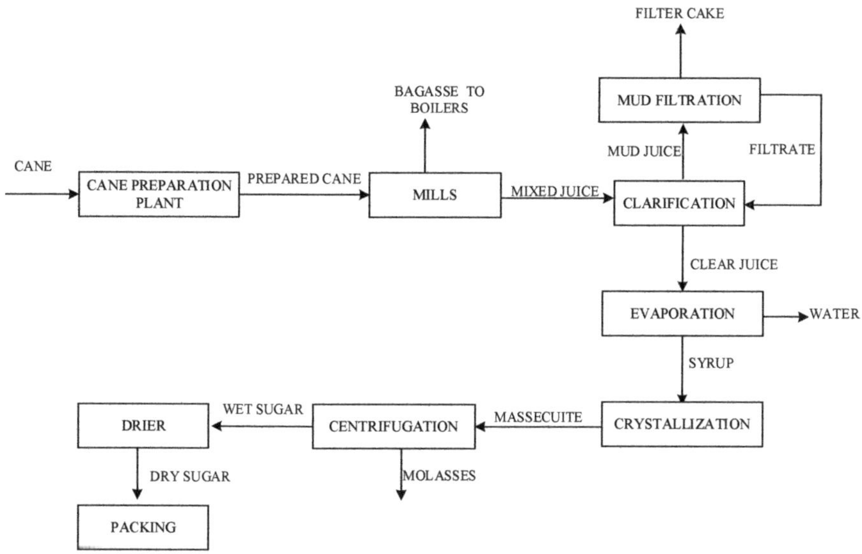

Figure 1 sugar production process flow diagram

Total of sugar loss encountered by any factory during particular crushing season is one of the significant factors when assessing technical performance of that sugar factory. Sugar loss by operations of any sugar factory depends on performance in milling station, boiling house efficiency, machinery and skilled manpower. It has been reported that, for sugar plants installed before and after 1994 the operational loss of sugar should be kept below 2.10 pol% cane and 2.20 pol% cane respectively (Sharma Chetan and Kumar Vikas, 2015).

1.2 Background of the problem

Kagera Sugar Limited (KSL) as other conventional sugar factories, suffers the loss of sugar during sugar production. The sugar loss in operation from cane milling station to centrifugation station was examined and found to be 2.83 and 2.87 pol % cane for seasons 2015/16 and 2016/17 respectively (see appendix A). The total sugar loss by operational activities in KSL was assessed and compared to literally reported loss figures for cane sugar industries. The loss for both seasons was higher than 2.20 pol% cane for plants installed after 1994 as reported by (Sharma Chetan and Kumar Vikas, 2015), so there was a need to investigate and give possible solution.

1.3 Significance of the study

The author was confident to say, if the proposed methods to control operational losses are employed, KSL will not only improve technical performance of the plant but also will increase business profit and help the struggle towards meeting the country's sugar demand.

1.4 Methodology

This study was undertaken under method of quantitative and arm chair thinking type of research. The data was collected from KSL reports of annual production of seasons 2015/16 and 2016/17.

2. Results and Discussion

KSL as any other conventional sugar mill, the total operational loss of sugar is broken into number of specific components. Operational sugar loss experienced during sugar production process in KSL is classified into the following components:

2.1 Sugar loss in final molasses

Sugar loss in molasses is highly dependent on the efficiency of the process and cane quality, the only factor that can be manipulated to minimize the loss by molasses is optimization of a process. For conventional factories, Molasses gives a loss of 1 to 1.3 pol% cane (Sharma Chetan and Kumar Vikas, 2015). But for KSL, loss in molasses for seasons ending 2015/16 and 2016/17 were 1.43 and 1.48 pol % cane respectively (see appendix A). As shown in figure 3, loss in molasses made 50.53 and 51.63 % of the total loss in the respective seasons. This means that, more sugar was lost in molasses compared to other operational losses.

2.1.1 How is sugar lost in molasses?

Loss of sugar in molasses is the function of processing techniques of boiling house. As reported by (Love DJ et al., 2009), sucrose is lost through molasses because during separation of mother liquor and sugar crystals some sucrose will remain in the final molasses. The author reports that there is a lower limit of exhaustion below which the amount of sucrose in final molasses cannot be reduced, this is defined by solubility of sucrose in molasses. Also (Hugot, 1986) reports that mediocre molasses exhaustion gives molasses of purity of 40 while exceptional exhaustion gives a final molasses of purity of 28. This means there is a limited chance that final molasses purity drops below 28. Final molasses purity is therefore a function of processing technique, cane quality and geographical position of the cane sugar factory.

2.1.2 What to be done to minimize sugar loss in molasses

The research by (Love DJ et al., 2009) , reports that to minimize the loss in molasses three laws of molasses loss should be used:

First law: The quantity of non-sucrose in final molasses must be minimized because, at any given purity P, each unit of non-sucrose will take with it P/(100-P) units of sucrose. This means if P is constant, to reduce loss in final molasses, only the quantity of non-sucrose should be minimized. This can be done by improving clarification process.

Second law: The achievable final molasses purity or target purity must be kept as low as possible except where this would contradict the First Law. The only way to reduce the target purity of final molasses can be speculated from the equation of calculating target purity adopted from (Rein, 2007):

Target Purity (South Africa) = $43.1 - 17.5 *(1 - e^{-0.74 \cdot (F+G)/A})$

To reduce the target purity is only by increasing the factor (F+G)/A, as it can be seen on the plot hereunder:

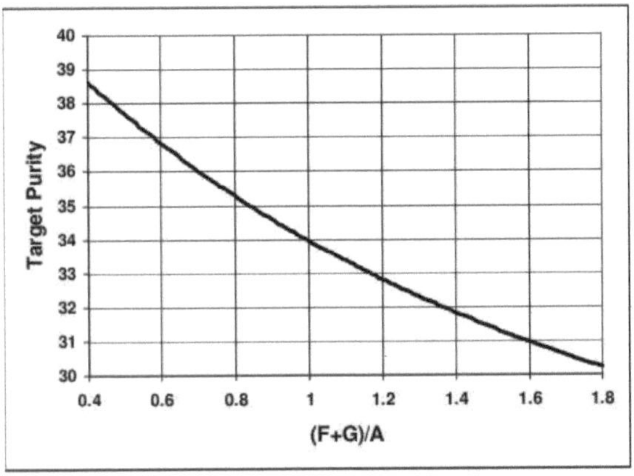

*Figure 2 The influence of (F+G)/A ratio on Target Purity (adopted from Love DJ et al., 2009) **F**- fructose% final molasses, **G**- glucose% final molasses and **A**-Ash % molasses*

This factor can be increased by adding the quantity of **F** and **G** and reducing **A,** but beware not to increase amount of F and G by inversion of sugar (which of course contradicts the first law). The only value to manipulate on this law is **A**, reduce the quantity of ash by proper clarification. Also Do not destroy the F+G to create other non-sucrose for the first law will also be contradicted.

Third law: The actual final molasses purity must be reduced to as close to the target purity as possible (or even better, below) except where this would also contradict the First or Second Law. This law deals with increasing exhaustion of final molasses purity to as close as target purity without increasing non-sucrose and/or reducing target purity. The author depicts that achieving low final molasses purity is a combination of optimized performances of seed preparation station, massecuite boiling, massecuite reheating, massecuite centrifugation.

All the laws, especially the third law if well implemented will help KSL produce much sugar with less loss in final molasses. And since molasses loss is the greater of all losses, it would also help reduce total seasonal sugar loss.

2.2 Sugar loss in bagasse

The aim of the milling station is to extract more sugar into mixed juice with minimum loss to bagasse. For good extraction sugar (pol)% bagasse from the last mill should be <1% (0.3 pol% cane) (Illovo Sugar limited (TOPs)., 2001). It's this pol% bagasse which will determine how much sugar is lost in bagasse in a particular crush year. For KSL the target loss in bagasse is <3 pol% final bagasse (0.9 pol% cane) which is relatively a higher figure comparing to that reported literally by (Illovo Sugar limited (TOPs)., 2001). For seasons 2015/16 and 2016/17 the loss in bagasse in KSL was 1.16 and 1.18 pol % cane (see appendix A).

2.2.1 What to be done to minimize sugar loss in final bagasse

Since the loss in bagasse is a function of performance of extraction station. And that performance is assessed by pol% bagasse which is in the other way a function of moisture content. If other conditions are kept constant, reduction in moisture content of the bagasse results in higher extraction. Reduction in bagasse moisture by 1% results in increase in extraction by 0.1% (Payne, 1982). It's clear that, only way to minimize loss in bagasse is therefore to improve extraction station efficiency. Note that extraction performance depends on quality of cane (fiber content), mill efficiency and efficiency of cane preparation plant in terms of preparation index (P.I), imbibition water and mill hydraulic load. The down jotted methods can be used to improve extraction station performance in KSL:

a) **Improve the P.I value**: this parameter should not be too low because it signifies the percent of cane cells opened for sugar extraction. However, it also should not be too high because the cane will be finely chopped leading to dropping at the mill and therefore poor extraction (Illovo Sugar limited (TOPs)., 2001). The practical value of this factor for good extraction is between 90-92%. For milling tandem, the value of 92% is impractical as cane will be too finely chopped therefore optimum value of 85% is satisfactory. As reported by (Rein, 2007), P.I increased from the value below 68% to 87% will result in extraction of 94.5% by so doing the loss in bagasse will be reduced compared to the current target extraction of 92% of KSL.
b) **Cane quality:** Cane with maximum extraneous matter gives lower extraction compared to the one with lower extraneous matter content. Fiber content of cane also affects mill setting and therefore extraction (Illovo Sugar limited (TOPs)., 2001). Cane with higher fiber content gives lower extraction and vice versa. KSL should encourage supply of cane which is clean and fresh to improve extraction.

c) **Mills efficiency**: mills should be set properly in accordance with the type of cane to be crushed in the particular crushing season and required fixed Tons cane crushed per hour (TCH). This will reduce the problem of high bagasse moisture due to varied throughputs and unstable mills speed. The mills also should be equipped with required hydraulic load and proper sanitation system to prevent microbial growth.

d) **Imbibition:** Too low imbibition will lower extraction and too higher imbibition will increase energy consumption of the plant in milling and evaporation station. For a milling tandem, extraction increases rapidly as imbibition increases from 250-280%. Imbibition above 280% on fiber, rate of response on extraction decreases the optimum value of 350% on fiber gives a significant gain in extraction (Rein, 2007). KSL should adopt this figure to minimize loss of sucrose in bagasse. Also the water of imbibition should be applied at maximum possible temperature not to cause feed ability problem (Payne, 1982). If feed ability problem due to high imbibition temperature exists mill arcing is inevitable. However, optimum temperature for imbibition water is 80^0C as reported by (Hugot, 1986)

2.3 Sugar loss in filter cake

Sugar lost in filter cake can also be estimated from the parameter pol% filter cake. For a conventional plant, this parameter should be <1% (Payne, 1982). The loss in bagasse can also be kept between 0.05- 0.15 pol % cane. For seasons 2015/16 and 2016/17 the loss in filter cake was maintained at 0.05 pol % cane which means KSL is doing a good work in maintaining low loss in bagasse.

2.3.1 What to do to further minimize sucrose loss in filter cake

- Ensure good and effective washing of filter cake during filtration, temperature of wash water should be 88-90°C and amount of about 100-150 % cake. The nozzle pressure should be 3-4 kg/m^2 (0.03-0.04kPa) (Hugot, 1986)
- Check and monitor vacuum control system for mud filters if they adhere to standards of -20 to -40kPa (for low vacuum) and -53 to -67kPa (for higher vacuum) (Hugot, 1986)
- Reduce the rotational speed of drum of mud filter if necessary (recommended value is 0.2 rpm) (Hugot, 1986).
- Regular check of cake thickness on the screen and reduce it according to regular filtration procedure, check for defective screens and replace.

2.4 Sugar loss in undetermined channels:

The undetermined loss in KSL are mainly due to mechanical loss, chemical causes and administrative loss. These losses for two previous seasons (2015/16 and 2016/17) were 0.19 and 0.17 pol % cane.

The losses are relatively high since, some factories are supposed to keep it between 0.01 to 0.11 pol % cane (Sharma Chetan and Kumar Vikas, 2015). To improve this KSL technical team should ensure all the mechanical leaks in the pumps, flanges and pipes are arrested. On the other hand, the production team should make sure all the production parameters are adhered to, to prevent loss by chemical means. Furthermore, the accuracy of weighing instruments should be regularly checked to prevent administrative loss.

3. General conclusion

3.1 From graph in figure 3, out of all loss, about 90 % of the total loss is contributed by molasses and bagasse loss. This means by using Pareto analysis, molasses loss and bagasse loss should be dealt with first. The use of three laws of sucrose loss in molasses coupled with optimization of boiling house operations would result in minimized loss of sucrose in molasses. Also, improving extraction would result in minimum sucrose loss in bagasse of possible solutions provided in this paper.

3.2 The company (KSL) should encourage the use of true purity instead of apparent purities because it's difficult to estimate effectiveness of clarification and molasses exhaustion in the course of cane juice processing. And therefore lead to failure to consistency during cane processing.

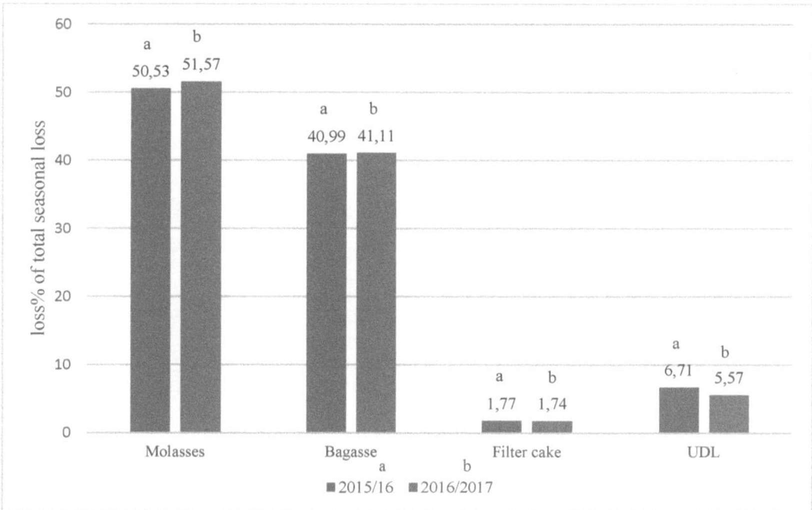

Figure 3 Contribution of loss in the total loss by individual loss channels in KSL

Appendix A

1. Production reports

Table 1 season 2015/16 production report

PRODUCTION REPORT FOR SEASON 2015/16			
CANE CRUSHED (TONS)	623256		
SUGAR BAGGED (TONS)	60988.5		
		Tons Pol	Loss sucrose % cane
Pol% cane	12.53	78093.98	
Pol molasses % pol in cane	11.43	8926.14	1.43
Pol bagasse % pol in cane	9.29	7254.93	1.16
Pol filter cake % pol in cane	0.42	327.99	0.05
Pol UDL % pol in cane	1.53	1194.84	0.19
Total	-	**17703.9**	**2.83**

PRODUCTION REPORT FOR SEASON 2016/17			
CANE CRUSHED (TONS)	665240		
SUGAR BAGGED (TONS)	67370.8		
		Tons Pol	Loss pol % cane
Pol% cane	12.91	85882.48	
Pol molasses % pol in cane	11.47	9850.72	1.48
Pol bagasse % pol in cane	9.14	7849.66	1.18
Pol filter cake % pol in cane	0.37	317.77	0.05
Pol UDL % pol in cane	1.26	1082.12	0.16
Total loss	-	**19100.3**	**2.87**

Table 2 season 2016/17 production report

PRODUCTION REPORT FOR SEASON 2016/17			
CANE CRUSHED (TONS)	665240		
SUGAR BAGGED (TONS)	67370.8		
		Tons Pol	Loss pol % cane
Pol% cane	12.91	85882.48	
Pol molasses % pol in cane	11.47	9850.72	1.48
Pol bagasse % pol in cane	9.14	7849.66	1.18
Pol filter cake % pol in cane	0.37	317.77	0.05
Pol UDL % pol in cane	1.26	1082.12	0.16
Total loss	-	**19100.3**	**2.87**

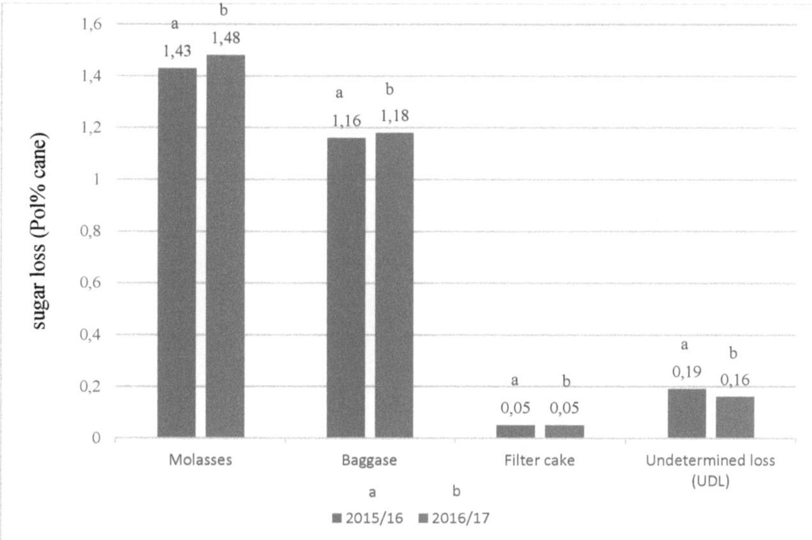

Figure 4 combined presentation of sugar loss (pol % cane) in KSL

References

Hugot, E. (1986). *Hand book of Cane Sugar Engineering.* Newyork: Elsivier.

Illovo Sugar limited (TOPs). (2001). *Technical Operating Practices (TOPs) : The Milling process.* Mount Edgecombe: Illovo Sugar Limited.

Love DJ et al. (2009). Minimizing Sucrose loss in Final Molasses : The three laws of Molasses loss. *Proc S Afr Sug Technol Ass*, 319-330.

Payne, J. H. (1982). *Unit Operation in Cane Sugar Production.* Newyork: Elsevier Scientific Publishing Company.

Perk, C. G. (1973). *The Manufacture of Sugar from Sugar cane.* Durban, South Africa: HAYNE & GIBSON LIMITED.

Rein, P. (2007). *Cane Sugar Engineering.* Berlin: Baton Rouge, Lousiana.

Sharma Chetan and Kumar Vikas. (2015, June). Quantification of Sugar Content Losses in various Byproducts of the Sugar Industry. *International Journal Of Advance Industrial Engineering, 3*, 50-51.

SMRI. (2012). *Essential Cane Sugar Technology.* Durban: SMRI South Africa.

YOUR KNOWLEDGE HAS VALUE

- We will publish your bachelor's and master's thesis, essays and papers
- Your own eBook and book - sold worldwide in all relevant shops
- Earn money with each sale

Upload your text at www.GRIN.com and publish for free

GRIN